YOUR KNOWLEDGE HAS VALUE

- We will publish your bachelor's and master's thesis, essays and papers

- Your own eBook and book - sold worldwide in all relevant shops

- Earn money with each sale

Upload your text at www.GRIN.com
and publish for free

T.S. Amar Anand Rao

New insights on Winogradsky Columns: Simulation of Contaminated Subsurface Systems for Low Cost, Sustainable Bioremediation

GRIN Verlag

Bibliografische Information der Deutschen Nationalbibliothek:

Die Deutsche Bibliothek verzeichnet diese Publikation in der Deutschen National-
bibliografie; detaillierte bibliografische Daten sind im Internet über http://dnb.d-
nb.de/ abrufbar.

Imprint:

Copyright © 2011 GRIN Verlag GmbH
Druck und Bindung: Books on Demand GmbH, Norderstedt Germany
ISBN: 978-3-656-09119-6

New insights on Winogradsky Columns: Simulation of Contaminated Subsurface Systems for Low Cost, Sustainable Bioremediation

T.S. Amar Anand Rao

Molecular Biophysics Unit, Indian Institute of Science

Bangalore – 560012, Karnataka, India.

Abstract

Everything is everywhere. The environment selects. Practical applications possible out of winogradsky column like using it as a universal enrichment medium for all microbes to grow as they are and also to isolate and evolve purpose based microbes for degradation studies. Nature is just a hidden Winogradsky column. Thus whatever mechanisms that underlies degradation in field is the same when simulated inside a winogradsky column. So we simulate all the constraints that face the degradation of hard substrates and the mechanisms inside this column to achieve results that can be immediately applied in field, without need of elaborate experimentation with artificial culture medium.

Keywords

Winogradsky Column/ Microcosm/ Biofilm/ microbial succession/ simulation/ degradation/ nutrient cycle

Introduction

The Winogradsky column is a classic demonstration of the metabolic diversity of prokaryotes. All life on earth can be categorised in terms of the organism's carbon and energy source: energy can be obtained from light reactions (phototrophs) or from chemical oxidations (of organic or inorganic substances) (chemotrophs); the carbon for cellular synthesis can be obtained from CO2 (autotrophs) or from preformed organic compounds (heterotrophs). Combining these categories, we get the four basic life strategies: photoautotrophs (e.g. plants), chemoheterotrophs (e.g. animals, fungi), photoheterotrophs and chemoautotrophs. Only in the bacteria - and among the bacteria within a single Winogradsky column - do we find all four basic life strategies.

The idea of a winogradsky column is this. In the foot or so of the column mud, with a little standing water on top slightly open to the air, two gradients appear. The mud is oxygenated at the top, but oxygen falls off as you go deeper, and an opposite sulfur gradient appears (not sure why, maybe due to the action of the bacteria?), high sulfur at the bottom which decreases as you go up the column. This provides a series of horizontal zones where various types of photosynthetic bacteria (& various other microbes) can find the oxygen and sulfur concentrations they like.

The references say (full Z) cynobacteria grow in the water. In the mud aerobic photosynthetic bacteria grow near the top (ripping apart water and outputting oxygen), while down near the bottom anaerobic photosynthetic bacteria grow, typically green sulfur and purple sulfur bacteria. These bacteria can fix carbon, but they rip apart sulfur compounds instead of water and output sulfur.

It's sometimes claimed a winogradsky column is like a section of a pond bed, but that cant' be right, because here the outside of the mud column, even a foot down where there is little oxygen and lots of sulfur, is bathed in light. The photosynthetic bacteria respond by growing in the illuminated outside layer of mud. (http://twinkle_toes_engineering.home.comcast.net/photosynthesis.htm)

Viruses are powerful manipulators of microbial diversity, biogeochemistry, and evolution in the marine environment. Viruses can directly influence the genetic capabilities and the fitness of their hosts through the use of fitness factors and through horizontal gene transfer. However, the impact of viruses on microbial ecology and evolution is often overlooked in studies of the deep subsurface biosphere. Subsurface habitats connected to hydrothermal vent systems are characterized by constant fluid flux, dynamic environmental variability, and high microbial diversity. In such conditions, high adaptability would be an evolutionary asset, and the potential for frequent host—

virus interactions would be high, increasing the likelihood that cellular hosts could acquire novel functions. Here, we review evidence supporting this hypothesis, including data indicating that microbial communities in subsurface hydrothermal fluids are exposed to a high rate of viral infection, as well as viral metagenomic data suggesting that the vent viral assemblage is particularly enriched in genes that facilitate horizontal gene transfer and host adaptability. Therefore, viruses are likely to play a crucial role in facilitating adaptability to the extreme conditions of these regions of the deep subsurface biosphere. We also discuss how these results might apply to other regions of the deep subsurface, where the nature of virus–host interactions would be altered, but possibly no less important, compared to more energetic hydrothermal systems.(Rika E. Anderson, 2011)

Chemical Processes within the Microbial Mat Community:

Cyanobacteria: $6 CO_2 + 6H_2O \rightarrow 6CH_{12}O> +6O_2$

Aerobic Heterotrophs: $6CH_2O> + 6O_2 \rightarrow 6CO_2 + 6H_2O$

Chemolithotrophic Bacteria: $H_2S + 2O_2 \rightarrow H_2SO_4 + ENERGY$

Phototrophic Bacteria: $4CO_2 + 2H_2S + 4H_2O \rightarrow 4CH_2O> + 2H_2SO_4$

Fermenters: $6CH_2O> \rightarrow 2C_2H_5OH + 2CO_2$

Sulfate Reducers: $SO_4^{2-} + CH_3COOH \rightarrow HS- + 2HCO_3$

Methanogens: $CH_2COOH \rightarrow CH_4CO_2$

$CO_2 + 4H_2 \rightarrow CH_4 + 2H_2O$ (© 2011 by the Yale-New Haven Teachers Institute)

The Daliao River, as an important water system in Northeast China, was reported to be heavily polluted by polycyclic aromatic hydrocarbons (PAHs). Aerobic biodegradations of four selected PAHs (naphthalene, phenanthrene, fluorene and anthracene) alone or in their mixture in river sediments from the Daliao River water systems were studied in microcosm systems. Effects of additional carbon source, inorganic nitrogen and phosphorus, temperature variation on PAHs degradation were also investigated. Results showed that the degradation of phenanthrene in water alone system was faster than that in water-sediment combined system. Degradation of phenanthrene in sediment was enhanced by adding yeast extract and ammonium, but retarded by adding sodium acetate and not significantly influenced by adding phosphate. Although PAHs could also be biodegraded in sediment under low temperature (5°C), much lower degradation rate was observed. Sediments from the three main streams of the Daliao River water system (the Hun River, the Taizi River and the Daliao River) demonstrated different degradation capacities and patterns to four PAHs. Average removal rates (15 or 19 d) of naphthalene, phenanthrene, fluorene and anthracene by sediment were in the range of 0.062–0.087, 0.005–0.066, 0.008–0.016 and 0–0.059 mg/(L·d),

respectively. As a result, naphthalene was most easily degraded compound, anthracene was the hardest one. In multiple PAHs systems, the interactions between PAHs influenced each PAH biodegradation.(Xiangchun Quan et al., 2009)

There are many PAH-degrading bacteria in mangrove sediments and in order to explore their degradation potential, surface sediment samples were collected from a mangrove area in Fugong, Longhai, Fujian Province of China. A total of 53 strains of PAH-degrading bacteria were isolated from the mangrove sediments, consisting of 14 strains of phenanthrene (Phe), 13 strains of pyrene (Pyr), 13 strains of benzo[a]pyrene (Bap) and 13 strains of mixed PAH (Phe + Pyr + Bap)-degrading bacteria. All of the individual colonies were identified by 16S rDNA sequencing. Based on the information of bacterial PCR-DGGE profiles obtained during enrichment batch culture, Phe, Pyr, Bap and mixed PAH-degrading consortia consisted of F1, F2, F3, F4 and F15 strains, B1, B3, B6, B7 and B13 strains, P1, P2, P3, P5 and P7 strains, M1, M2, M4, M12 and M13 strains, respectively. In addition, the degradation ability of these consortia was also determined. The results showed that both Phe and mixed PAH-degrading consortia had the highest ability to degrade the Phe in a liquid medium, with more than 91% being degraded in 3 days. But the biodegradation percentages of Pyr by Pyr-degrading consortium and Bap by Bap-degrading consortium were relatively lower than that of the Phe-degrading consortium. These results suggested that a higher degradation of PAHs depended on both the bacterial consortium present and the type of PAH compound. Moreover, using the bacterial community structure analysis method, where the consortia consist of different PAH-degrading bacteria, the information from the PCR-DGGE profiles could be used in the bioremediation of PAHs in the future. (Tian Yun et al., 2010)

The vertical distribution of polycyclic aromatic hydrocarbons (PAHs) at different sediment depths, namely 0–2 cm, 2–4 cm, 4–6 cm, 6–10 cm, 10–15 cm and 15–20 cm, in one of the most contaminated mangrove swamps, Ma Wan, Hong Kong was investigated. It was the first time to study the intrinsic potential of deep sediment to biodegrade PAHs under anaerobic conditions and the abundance of electron acceptors in sediment for anaerobic degradation. Results showed that the total PAHs concentrations (summation of 16 US EPA priority PAHs) increased with sediment depth. The lowest concentration (about 1300 ng g− 1 freeze-dried sediment) and the highest value (around 5000 ng g− 1 freeze-dried sediment) were found in the surface layer (0–2 cm) and deeper layer (10–15 cm), respectively. The percentage of high molecular weight (HMW) PAHs (4 to 6 rings) to total PAHs was more than 89% at all sediment depths. The ratio of phenanthrene to anthracene was less than 10 while fluoranthene to pyrene was around 1. Negative redox potentials (Eh) were recorded in all of the sediment samples, ranging from − 170 to − 200 mv, with a sharp decrease at a depth of 6

cm then declined slowly to 20 cm. The results suggested that HMW PAHs originated from diesel-powered fishing vessels and were mainly accumulated in deep anaerobic sediments. Among the electron acceptors commonly used by anaerobic bacteria, sulfate was the most dominant, followed by iron(III), nitrate and manganese(IV) was the least. Their concentrations also decreased with sediment depth. The population size of total anaerobic heterotrophic bacteria increased with sediment depth, reaching the peak number in the middle layer (4–6 cm). In contrast, the aerobic heterotrophic bacterial count decreased with sediment depth. It was the first time to apply a modified electron transport system (ETS) method to evaluate the bacterial activities in the fresh sediment under PAH stress. The vertical drop of the ETS activity suggested that the indigenous bacteria were still active in the anaerobic sediment layer contaminated with PAHs. The biodegradation experiment further proved that the sediment collected at a depth of 10–15 cm harbored anaerobic PAH-degrading bacterial strains (two Sphingomonas, one Microbacterium, one Rhodococcus and two unknown species) with some intrinsic potential to degrade mixed PAHs consisting of fluorene, phenanthrene, fluoranthene and pyrene under low oxygen (2% O2) and non-oxygen (0% O2) conditions. This is the first paper to report the anaerobic PAH-degrading bacteria isolated from subsurface mangrove sediment.(Nora Fung-Yee Tam, 2009)

Mono-alkyl phthalate esters (MPEs) are primary metabolites of di-alkyl phthalate esters (DPEs), a family of industrial chemicals widely used in the production of soft polyvinyl chloride and a large range of other products. To better understand the long term fate of DPEs in the environment, we measured the biodegradation kinetics of eight MPEs (-ethyl, -n-butyl, -benzyl, -i-hexyl, -2-ethyl-hexyl, -n-octyl, -i-nonyl, and -i-decyl monoesters) in marine and freshwater sediments collected from three locations in the Greater Vancouver area. After a lag period in which no apparent biodegradation occurred, all MPEs tested showed degradation rates in both marine and freshwater sediments at 22 °C with half-lives ranging between 16 and 39 h. Half-lives increased approximately 8-fold in incubations performed at 5 °C. Biodegradation rates did not differ between marine and freshwater sediments. Half-lives did not show a relationship with increasing alkyl chain length. We conclude that MPEs can be quickly degraded in natural sediments and that the similarity in MPE degradation kinetics among sediment types suggests a wide occurrence of nonspecific esterases in microorganisms from various locations, as has been reported previously.(Frank A.P.C. Gobas, et al, 2008)

Aerobic biodegradation has been considered to be the main attenuation mechanism for microcystins, but the role of anoxic biodegradation remains unclear. We investigated the potential for anoxic biodegradation of microcystin and the effects of environmental factors on the process

through a series of well-controlled microcosm experiments using lake sediments as inocula. Microcystin LR could be degraded anoxically from 5 mg L−1 to below the detection limit at 25 °C within 2 days after a lag phase of 2 days. The rate was highly dependent on temperature, with a favorable temperature range of 20–30 °C. The addition of glucose or low levels of NH4-N had no effect on the anoxic biodegradation of microcystin, whereas the addition of NO3-N significantly inhibited the biodegradation at all experimental concentrations, and the inhibition increased with increasing amount of NO3-N-amended. Adda (3-amino-9-methoxy-2,6,8-trimethyl-10-phenyl-deca-4,6-dienoic acid), a previously reported nontoxic product of aerobic degradation of microcystin, was identified as the anoxic biodegradation product. This is the first report of Adda as a degradation product of microcystin under anoxic conditions. No other product containing Adda residue was detected during the anoxic degradation of microcystin. These results strongly indicated that anoxic biodegradation is an effective removal pathway of microcystin in lake sediments, and represents a significant bioremediation potential. (Bangding Xiao et al., 2009)

To contain domestic waste and its associated pollution within a landfill, engineered mineral (clay) barriers are used and are designed to have a permeability of 1×10^{-9} m/s. The rate of permeability of various porous media has shown to be influenced by the clogging of flow paths (media pores) due to biofilm formation. The term biofilm is given to describe the colonies of surface adherent microorganisms. In this study, permeability experiments were built and modified to act as microcosms to investigate the influence of biofilm formation on the permeability of clay barriers. Traditional scanning electron microscopy methods disrupt or destroy the biofilm and previous anaerobic studies have involved building closed cells (such as miniature continuous culture chambers) that utilise light microscopes. This paper examines the application of the environmental scanning electron microscope (ESEM) to the direct examination of the clay interface and biofilm formation in situ within the microcosm. (M. G. Darkin et al., 2001)

Materials and Methods

Pond soil and water were collected from the water soil interphase of ponds at Government Botanical Gardens, Ooty. This place was selected because these gardens have been undisturbed natural biosphere Procedure of making a Winogradsky column (Anderson et al 1999) the soil sample was cleaned of debris, stones, pebbles, grass clippings, leaves and moving insects. This is used as the Control Column A (Figure1) for all the other variations.

Fill one fourth of a 250ml glass measuring cylinder with the soil. Mix the soil with 2g of cellulose, 2g of calcium carbonate, and 2g of calcium sulphate. Cover upto three fourth of the column with soil slurry. Let the soil set for five minutes to release trapped air bubbles. Add water leaving a 2cm gap at top. Incubate the column where it will receive daylight or artificial light. Observe the column over the next several weeks for development of layers, smell, colours, and zones.

Simulation of varied environmental factors
By building variations of Winogradsky column one factor at a time- other conditions were kept unchanged.

The hard substrate that is to be degraded can be layered at the bottom of the column or as alternative layers. (Dorothy M May, 1991)

Variations of nutrients/starvation of nutrients
Column 1 contains carbon source only: 2 g cellulose , Column 2 contains sulphur source only: 2g $CaSO_4$, Column 3 contains carbonate source only: 2g of $CaCO_3$, Column 4 has no added nutrients at all (Figure2)

Variations of pH
pH values were maintained by adding buffer tablets to the water of the column. Column 5 has pH3 . Column has 6 pH5 . Column 7 has pH7 . Column 8 has pH9(Figure3).

Variations of salinity
Winogradsky column was watered with the following values of salinity of water. It decides the osmotic pressure. Column 9 has 1% salt concentration. Column 10 had 2.5% salt concentration , Column 11 has 5% salt concentration. Column 12 has 10% salt concentration(Figure 4)

Variations of texture

The texture of the soil decides the porosity. Column 13 has red garden soil. Column 14 has sponge. Column 15 has sand. Column 16 has clay.(Figure 5)

Variations of hard substrates

To make degradation possible, the columns were made to contain the hard substrates as a sole source of carbon by starvation. Column 17 has coir . Column 18 has dye. Column 19 has naphthalein PAH . Column 20 has urea . Column 21 has paraffin . Column 22 has ash. Column 23 has chitin (Figure6 & 7).

Electrochemical gradient potential:

Iron corrosion is studied by using red mud high in ferrous content. In column 24 (Figure 8), the electrochemical gradient potential between the top and the bottom of the standard column was monitored by inserting multimeter probes (Anderson, 1999)

Isolation of Methylotrophs

Purpose: We may use methanol as the Sole source of carbon in Column 25 (Figure 9).

Incubation

Incubation was done at room temperature and under artificial lighting in algal growth chambers for three months. Readings were taken at the end of each month.

Tracking biofilm pattern

The changing biofilm patterns of the Winogradsky columns were kept track by tracing the outlines of the biofilm patterns with colour markers on a rectangular piece of polythene sheet around the column (NASA quest)

Quantitative data of biofilm patterns

The biofilm patterns out on polythene sheets were cut according to the color markers and weighed on a microbalance. Thus the equivalent weights of biofilm patterns were got (NASA quest).

We suggest an alternative which is very much realistic and useful. Download the panorama creating photo edit software if using linux (Ubuntu) like Hugin Panorama Creator (Figure 10), which is available as a free package in Ubuntu Software Center. This can stitch pictures taken from a single location using control points. The idea is to fix the Digital camera with a tripod and turn the

winogradsky in all four directions keeping it on a marked point on the table. This data can then be used to calculate area of each coloured visible biofilm pattern and used as a value for drawing graph at different time points. So now, a map of the whole surface of the Winogradsky column can be made with great accuracy. One more alternative is to download GIMP Image processing software and add the Pandora plugin. This is even better in creating panorama images to create Winogradsky maps as JPEG files.

For calculation of area, Image J Image processing and analysis software (Figure11) is available in Ubuntu for free. It can calculate area and pixel value statistics of user defined selections. The JPEG image obtained from the Panorama software can be used here to calculate the biofilm area.

Creating the Biogeochemical Map (Figure 12)
A detailed Biogeochemical cycling pathways based on all carbon and energy flux happening in the micro and macroenvironments of organic and inorganic contaminated subsurface systems is made. To make this possible all possible factors affecting are first listed or atleast the most important ones. And correlate all the pathways to find out where the hard substrate in the given locality stands. This will help to simulate and build better Winogradsky to test and implement the induction of stress to adapt or evolve new microbes to degrade the hard substance.

Results

Effects of environmental factor variation on biofilm pattern and isolation of adapted populations

Simulation of microbial succession

Microbial populations grew in succession in the winogradsky column kept as standard. A white biofilm at the bottom at the bottom followed by black coloration of soil followed by purple and green biofilms appeared in the soil layer with green algal biofilms in the water layer.

Variants of the winogradsky column gave different biofilm patterns indicating that different pond microcosms can be simulated by manipulating the environmental factors. So it is worth designing a grand column where we can manipulate all factors precisely. All results of column variations were interpreted using table. The change of patterns and the appearance and disappearance of patterns indicates the changes occuring in the microenvironments created by gradients developed according to the factors manipulated.

Nutrients/ Starvation

Even in limiting conditions of nutrients, green, rust, and black biofilm patterns appeared indicating their ability to survive in such conditions. Which may also mean that soil by itself has enough carbon substrates to be degraded anaerobically and start the microbial succession. Though the variety of biofilms and their density are poor due to lack of all the nutrients. But it shows that the surviving ones are highly adapted to survival under stressed environments. Starvation induces the selective pressure needed to select those adapted microbes and enrich them to form visible patterns. These can be isolated using a pipette. Starvation of nutrients is the best method to evolve and adapt microbes to degrade them.

pH

All biofilm patterns decreased with increase in pH indicating their ability to tolerate low pH better. The H_2S production increases drastically with increase in pH. And in low pH the microbes isolated or biofilms visible are only those in the microaerophilic region. The pH of the locality where the hard substrates are contaminated, plays a vital role in deciding the biodegradation of the substrate. Methane is produced in very high levels in higher pH. It displaces the mud and comes out in liters. This can be tested by piercing a air pocket in the winogradsky with a pipette and holding a lighted match stick to burn the escaping methane. It also explains the nature of methane bogs.

Salinity

Biofilm patterns decreased with increase in salinity indicating intolerance only at high salinity. It is of significance because it simulates fresh water river system degradation pattern and saline water sea degradation patterns. Especially this plays a critical role in selecting the algal forms in the water column. It gave a impressionable diversity of cyanobacteria, slimy filamentous filaments and mats of *Phormidium* sp. (Figure13)

Texture

Red soil and sand show good biofilm patterns indicating the respective porosities and texture of these soils to harbour most biofilm patterns. Different regions have differnent soil textures. This simulation shows obvious differences in pattern of degradation biofilms and microbe population respective to the texture of the soil of the locality contaminated with hard substrates. The porosity is decided by the soil texture and the aeration too depends on it. Sandy soil gives high aeration and is microaerophilic. While the clay soil has low porosity due to fine texture and gives patterns dark, resembling those found in peat bogs and is usually anaerobic.

Hard substrates

All hard substrate column showed slow gradual and dotted patterns of black and green biofilms in the anaerobic region, indicating evidence of the ability of biofilms to adapt and degrade any hard substrates (pollutants of human activity in water bodies) by the method of high frequency gene transfer based evolution that takes place only in biofilms. Starvation on a specific hard substrate allows new microbial populations to evolve, visible as dotted populations arising at the site of disappearing or degraded dye in the column. In nature we have to remove other carbon sources to degrade the hard substrate of choice. Anaerobic regions favor bioremediation as that is the place where the carbon flux starts. Additives like very minimum amounts of glucose can enrich degrading microbes to start off initially before adapting to the hard substrates. This is a important tool in degradation. Samples obtained from these columns can be made use to do study viral mediated gene transfer based evolution.   Naphthalene (PAH), Industrial Dyes and Coir were easily degraded as sole substates of carbon.

Electrochemical gradient potential

The electrochemical potential probed between the top and bottom of a column plate showed increase in value from 0 to 500mV with time from 0 to 90 days indicating the ability to produce tiny amounts of electricity. It may be negative redox potential and it increases with depth. Demonstrates that carbon electrodes stuck deep into a peat bog and the other at the surface can be used to harness

microbial redox electricity.

These were sufficient evidence that if a column is designed to have all minute capacities of variation of environmental factors to bring about the specific environment of sampling water body, a perfect simulation of the biofilm patterns can be achieved and the resulting microbes are highly adapted to survive in those specified conditions. And can be isolated.

Isolation of Methylotrophs : A test to isolate purpose based microbe using Winogradsky simulation. A range of methlylotrophs ranging from algae to bacteria were obtained. Pink Pigmented Facultative Methylotrophs were isolated in the water column and cultured using MRS medium. *Scenedesmus obliquus* unicellular methylotroph algae (Figure14) was isolated from the water column. By just altering the carbon source, we can get those specific adapted microbes suitable to degrade it. Below layers showed biofilm mats that had populations degrading 100 % methanol since the column was made from Methanol instead of water.

Biofilms were removed from the column after degradation took place to reveal the microbes(Figure15). But for quantitation purposes, digital imaging photography is better.

Taking into account of natural phenomenon taking place inside Winogradsky column
To make use of natural biogeochemical and microbial processess to harness their bioremdiation potential, a clear picture of the various useful phenomenon was listed to take into account to correlate with degradation studies. These are very much essential to cut down the cost and make biodegradation sustainable. And a through mapping of the biogeochemical cycles involved in these evolving degradation is necessary to make better simulations. (Figure 17)

Listing the phenomenons crucial for making the Biogeochemical Map based on induced carbon flux and energy flux pathways, useful for subsurface contaminated systems:-
Enrichment based degradation/ Selective pressure based degradation/ Additives based biodegradation/ Anoxic degradation/ Starvation based degradation/ Recalcitrant contaminant/ Organic sustances/ PAH/ Naphthalene/ Anthracene/ Auxillary metabolism based degradation/ Co-metabolism/ Carboxylation/ Gasolene additives/ Consortium of Mixotrophs degradation/ Microbial succession based degradation involving Phototrophs/ Organotrophs/ Lithotrophs/ Heterotrophs/ Chemolithotrophs/ Aerobes/ Microaerophiles/ Anaerobes/ Redox cycle based corrosion/ Altering Nutrients, pH, salinity, Texture, etc for degradation/ Carbon conversion (Figure16)

ImageJ (http://rsbweb.nih.gov)

ImageJ is written in Java, which allows it to run on Linux, Mac OS X and Windows, in both 32-bit and 64-bit modes. Open Source: ImageJ and its Java source code are freely available and in the public domain. No license is required.

Selections:

Create rectangular, elliptical or irregular area selections. Create line and point selections. Edit selectoins and automatically create them using the wand tool. Draw, fill, clear, filter or measure selections. Save selections and transer them to other images.

Analysis:

Measure area, mean, standard deviation, min and max of selection or entire image. Measure lengths and angles. Use real world measurement units such as millimeters. Calibrate using density standards. Generate histograms and profile plots.(http://rsbweb.nih.gov)

Any further analysis can be made using, Genius Mathematics Tool and the GEL Language (http://www.jirka.org/genius)

3D surface plot Genius is a general purpose calculator program similar in some aspects to BC, Matlab, Maple or Mathematica. It is useful both as a simple calculator and as a research or educational tool. The syntax is very intuitive and is designed to mimic how mathematics is usually written. GEL is the name of its extension language, it stands for Genius Extension Language, clever isn't it? In fact, many of the standard genius functions are written in GEL itself.

Legends for Pictures

Figure 1 Control Winogradsky column

Figure 2 Variants of nutrients

Figure 3 Variants of pH

Figure 4 Variants of salinity

Figure 5 Variants of Soil texture

Figure 6 & Figure 7 Variants of hard substrates

Figure 8 Redox potential

Figure 9 Absolute Methanol substrate

Figure 10 Panorama Hugin software in work finding control points to stitch two sides of a column

Figure 11 ImageJ software analysing area for a selected area from the panaroma view of column

Figure 12 Winogradsky carbon flux and microbial succession map

Figure 13 *Phormidium* sp. Cyanobacteria filaments

Figure 14 *Scenedesmus obliquus* Methylotrophic algae

Figure 15 Dioremediation biofilms layers removed from column

Figure 16 Biogeochemical cycles pathway map

Figure 17 A natural contaminated peat bog showing bioremediating biofilms

Figure 1

Figure 2

Figure 3

Figure 5

Figure 6

Figure 7

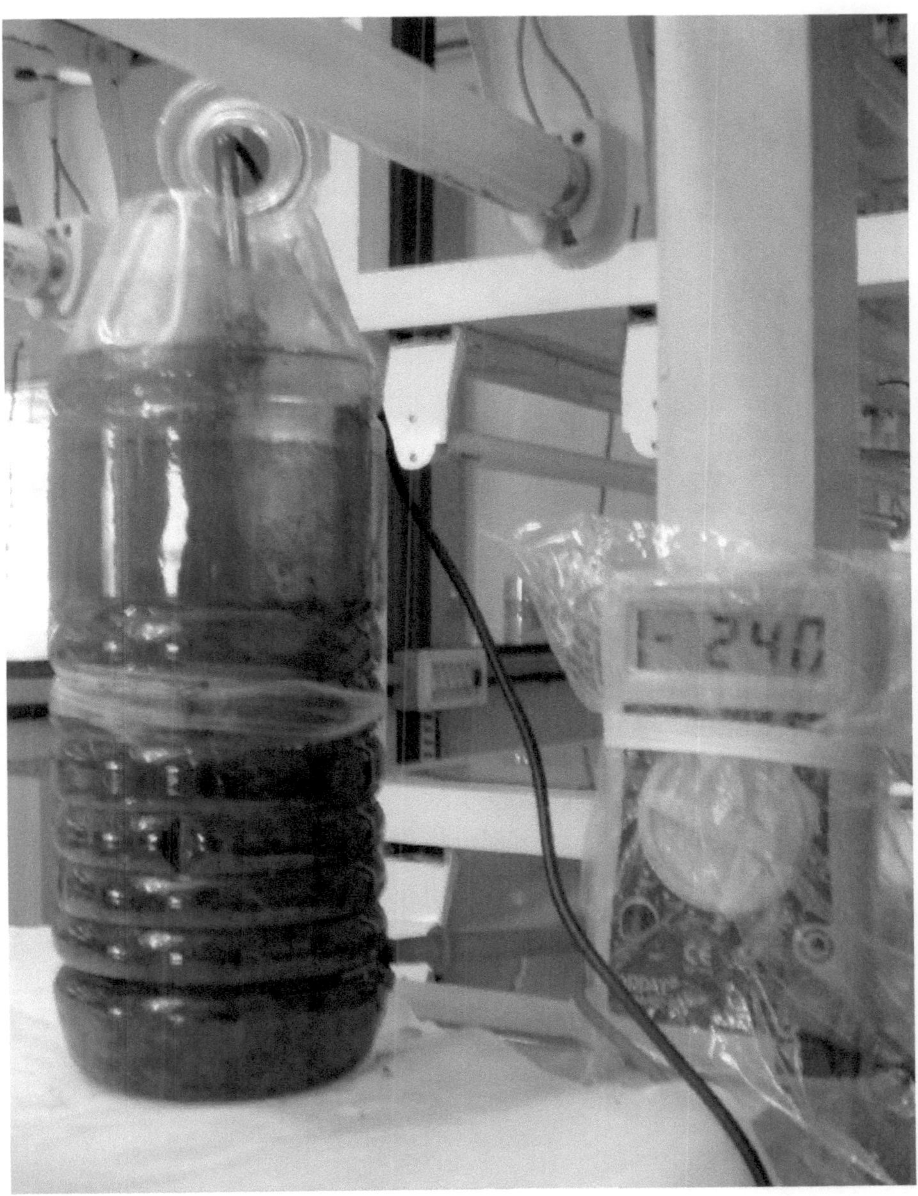

Figure 8

Figure 9

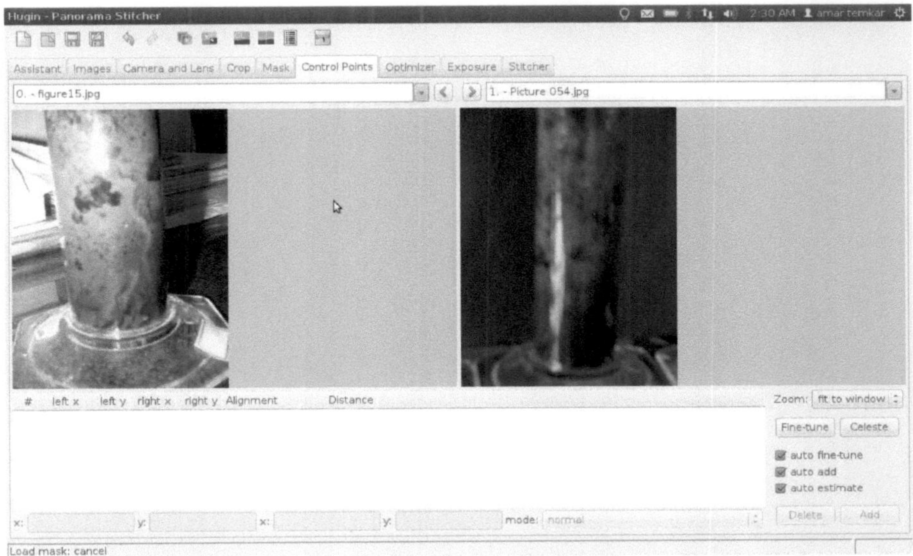

Figure 10

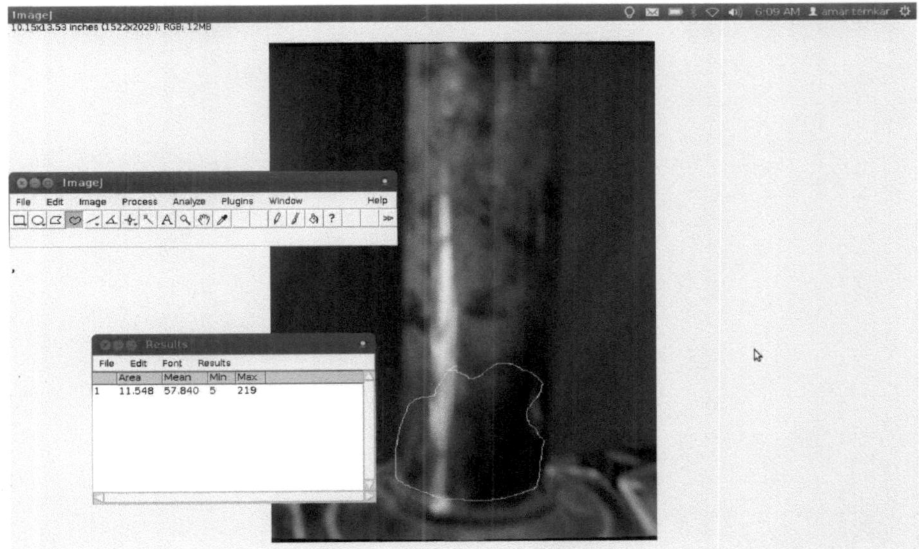

Figure 11

Idealized Version of a Microbial Mat

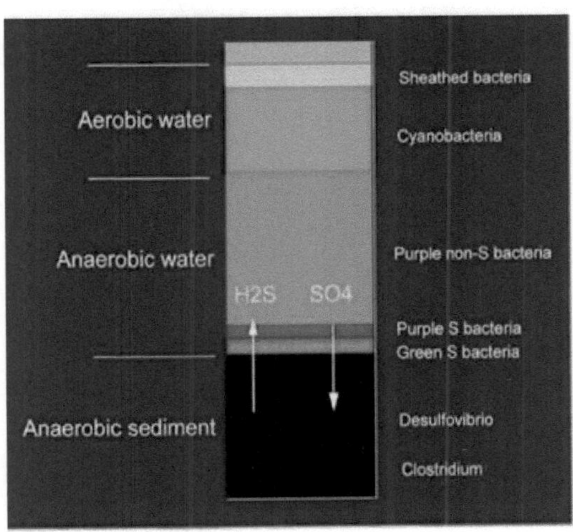

Figure 12 (source: http://twinkle_toes_engineering.home.comcast.net/photosynthesis.htm)

Figure 13

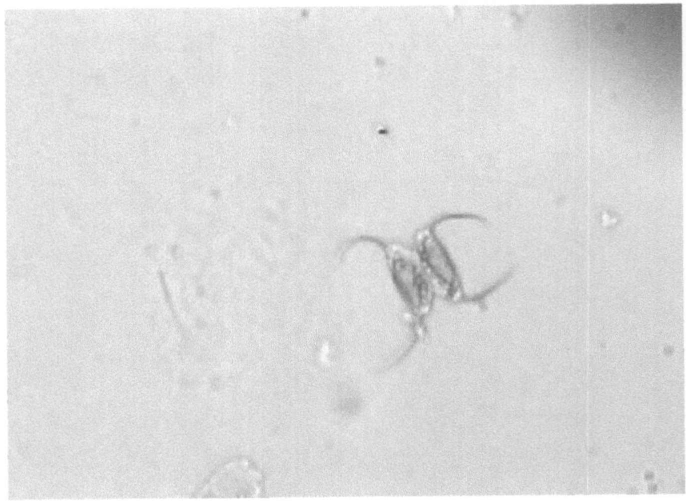

Figure 14

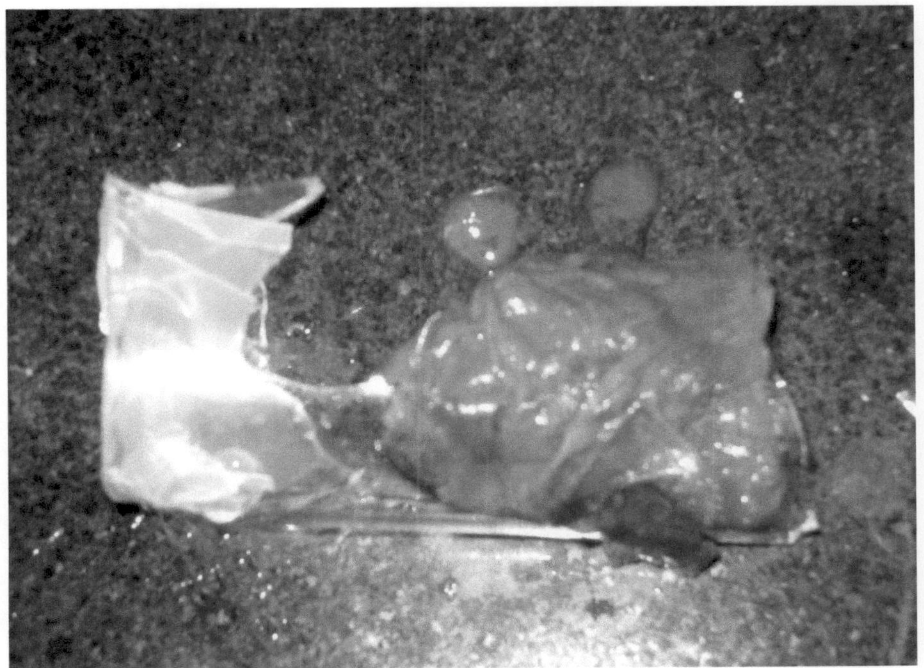

Figure 15

Microbial Mat Biogeochemical Cycling

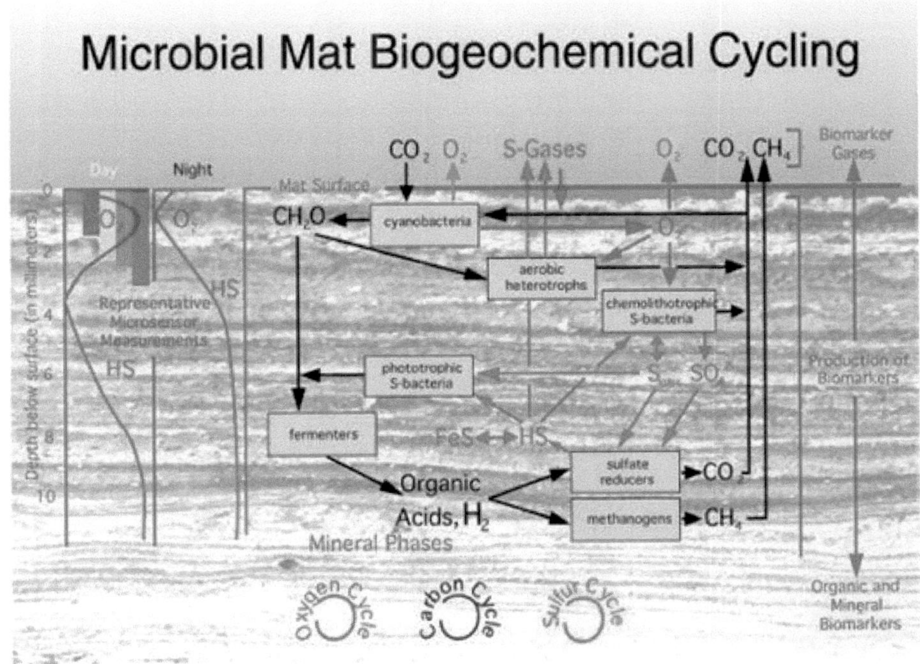

Figure 16 (Source: http://twinkle_toes_engineering.home.comcast.net/photosynthesis.htm)

Figure 17

Discussion

The basic inference of immense value got from this simulated winogradsky column experiments is the proof that microbial degradation is greatly dependent on the environment of the hard substrate contaminant. Using these studies provides a picture of what contribution each environmental factor plays in the degradation of the hard substrate. These factors thus understood can be employed in degradation of the contaminated area by altering the environment to suit the lab simulated conditions that show great degradation patterns. This allows for a consortium of microbial succession to degrade the hard substrate using carbon and energy fluxes through biogeochemical cycling.

This is a better way of degrading hard substrates naturally than relying on mutating a single biodegrading microbe to achieve the required result.

The tip is to harness the environmental energy flows of the contaminated area in such a way so as to favor the degradation of the contaminant through allowing the stress induced to evolve the microbes naturally present to adapt to the selective pressure.

So, to solve a contaminant problem in a locality lies in simulating it in a winogradsky to find the right set of combination to degrade and apply them to the contaminated subsurface system.

This is a cheap sustainable low cost method to degrade contaminants where they are, by inducing selective stress to evolve microflora. So that all the efforts needed can be shifted from hi end tools to simple practical biogeochemical microenvironment modification stations located at the very site of contamination till the degradation cycle starts. Then such valuble new highly evolved populations from these sites can be isolated and sequenced and put use directly elsewhere.

Quantitative studies can be correlated to the data obtained from biodegrading biofilm patterns to clearly understand and predict the behaviour and purpose of the microflora. This can better tell in statistical ways, how to proceed about in making contaminated subsurface systems to induce stress to the natural microflora in adapting or evolving new strains.

References

1. Xiangchun QUAN, Qian TANG, Mengchang HE, Zhifeng YANG, Chunye LIN, Wei GUO, Biodegradation of polycyclic aromatic hydrocarbons in sediments from the Daliao River watershed, China, Journal of Environmental Sciences, Volume 21, Issue 7, 2009, Pages 865-871

2. Liu HuiJiea, Yang CaiYuna, Tian Yuna, Lin GuangHuia, Zheng TianLinga, Using population dynamics analysis by DGGE to design the bacterial consortium isolated from mangrove sediments for biodegradation of PAHs, International Biodeterioration & Biodegradation, Volume 65, Issue 2, March 2011, Pages 269-275

3. Chun-Hua Lia, Hong-Wei Zhoub, Yuk-Shan Wongc, Nora Fung-Yee Tam, Vertical distribution and anaerobic biodegradation of polycyclic aromatic hydrocarbons in mangrove sediments in Hong Kong, South China, Science of The Total Environment, Volume 407, Issue 21, 15 October 2009, Pages 5772-5779

4.S. Victoria Ottona, Srinivas Surab, Joel Blairc, Michael G. Ikonomouc, Frank A.P.C. Gobasa, Biodegradation of mono-alkyl phthalate esters in natural sediments, Chemosphere, Volume 71, Issue 11, May 2008, Pages 2011-2016

5. Xiaoguo Chena, Xia Yanga, Lili Yanga, Bangding Xiaob, Xingqiang Wub, Jutao Wanga, Hongguo Wana, An effective pathway for the removal of microcystin LR via anoxic biodegradation in lake sediments, Water Research
Volume 44, Issue 6, March 2010, Pages 1884-1892

6. http://twinkle_toes_engineering.home.comcast.net/photosynthesis.htm

7. Rika E. Anderson, William J. Brazelton, and John A. Baross, Is the Genetic Landscape of the Deep Subsurface Biosphere Affected by Viruses?, Front Microbiol. 2011; 2: 219

8. Biodegradation © 2011 by the Yale-New Haven Teachers Institute

9. Dorothy M May, BIOSCENE, Vol.17 (3), December 1991, pages 10-11.

10. M. G. Darkin, C. Gilpin, J. B. Williams, C. M. Sangha, Direct wet surface imaging of an anaerobic biofilm by environmental scanning electron microscopy: Application to landfill clay liner

barriers, Volume 23, Issue 5, pages 346–350, September/October 2001

11. http://rsbweb.nih.gov
12. http://www.jirka.org/genius.html